A Brief Treatise on some Applications of Mnemotechnics

Olegario Nicasio

A Brief Treatise on some Applications of Mnemotechnics

Olegario Nicasio
First Edition

First edition

Author: Olegario Nicasio

Book design, graphics and illustrations: Olegario Nicasio

ISBN: 978-1-6887-7372-1

*Dedicated to all of those
who have been interested
in the art of memory.*

Contents

Prologue

The art of memory is an antique practice. *Ergo,* in this writing I don't pretend to do an exhaustive and rigorous work. Instead, I write the techniques that has served me better in my mnemotechnical experience. Although some of the things that work for me are subjective, I want to give the reader a glimpse to what works and what doesn't at the time of practicing the art of memory. This brief work is for the person who has an intermediate understanding of mnemotechnics. Consequently, I offer additional references for those who lack understanding of mnemotechnics or just want to broaden the view of mnemotecnics.

Also, I did this brief teatrise because some of my posts in mnemotechnics forums have been lost. Here I compile some of the core ideas I shared.

System for Number Encoding

1 Decimal Number System PAOT

The PAOT System [person, action, object and tool] is a variation of the Dominic PA System[1] [person and action]. The POAT or PATO system was developed by Octaveslur[2]. Hence, the PAOT is a variation I made using Octaveslur's system.[2] Below is a list of my images for numbers between 00-99. In most cases I use the Dominic System and sometimes I attribute an image to a number for what it means to me. So the images are not necessarily linked by a phonetic code.

Memorizing goes in this fashion: the number 45896327 is separated as 45-89-63-27. The number is memorized as "Dee-Dee [45] picking (like eating with a beak) [89] a hound [63] with a keyboard (the beak is a keyboard) [27]".

This is my list of images: Person	Action[3]	Object	Tool
01- Matt Damon	kicking	pole	karate suit
02- Samuel Jackson	screaming	Pulp Fiction Wallet	megaphone

---

[1] O'Brien, Dominic (1993). *How to Develop a Perfect Memory.* Headline Book Publishing.

[2] Octaveslur (2012). *Octaveslur's Digit System.* Recovered from http://mt.artofmemory.com/blogs/octaveslur/octaveslurs-digit-system-1889.html
[3] I have invented a lot of verbs that show the action I mean. Concurringly, the verbs are invented in function with the tool.

03- Kim Jong Il	watching	Mario power star	telescope
04- Joker	driving	car	pork chop
Belushi	putting	glove	blues suit
06- Aykroyd	firing	bullets	gun
07- Bayly	dicing	dice	stick
08- Coelho	spinning	skate	tutu
09- Dostoyesky	jumping	cat	pogo stick
10- Megan Fox	siting	bowling ball	pine
11- Nietzche	drumming	drum	drumstick
12- M.Jackson	Rubbing	dozen-eggs	"trapo"
13- Hipster-tiger	Roaring	witch	hipster-glasses
14- Thor	Striking	ring	lightning
15- Einstein	Writing	theory	pen
16- Schwarzenneger	Lifting	weights	shorts
17- Hawking	Reading	magazine	vox-machine
18- Talos	Beheading	priest	warhammer
19- The Thing	Breaking	bench	suit
20- Dragon Tattoo	smokes	cigar	cigarette-holder
21- Barbie	nailing	nails	Nails
22- Bilbo	dancing	fire	dragon-mail
23- Clinton	sucking	dick	mouth
24- Big Daddy	pissing	enchanter	cone
25- Uncle Ben	punching	throne	glove
26- Bart Simpson	skating	rails	skateboard
27- Bill Gates	computer	hacking	keyboard
28- Bounty Hunter	flying	jetpack	mask
29- Bono	singing	Kajit (Skyrim)	Nokia
30- Cops	beating	people	cop-sticks
31- Captain America	throwing	shield	C.A. logo
32- Batman	riding	unicorn	batman logo
33- Charlie Chaplin	balancing	tricycle	Hitler 'stache
34- Darwin	whispering	plant	recorder
35- Clint Eastwood	knotting	knots	knots
36- C.S. Lewis	splitting	Dawn Trader	force-field
37- Guevara	stomping	hat	das-boot (movie Beerfest)
38- Cho	hiding	treasure-chest	Digory
39- Chuck Norris	combing	beard	comb
40- Dominic	shuffling	cards	card shuffler

41- Daenerys	breast-feeding	dragon	breast
42- Goku	Kamehameha-wahing	dragonballs	Kamehameha
43- Ramona	dyeing	hair	blue-dye
44- Donald	quacking	book shelve	duck-mouth
45- Dee-Dee	perturbing distracting	dragon-priest	wrench
46- DS	slides	DS	water-slide
47- Dorian Grey	hammering	Dorian Grey	hammer
48- Hasselhoff	meeting	Hasselhoff	handshake
49- Barney	shoveling	dinosaur	shovel
50- Red Hot Chili Peppers	eating	peppers	peppers
51- Harry Potter	playing	videogames	controller
52- Mermaid Man	swimming	Barnacle Boy	fins
53- Eric Clapton	tuning	piano	tuner
54-Ellen Degeneres	flashing	flash	flashlight
55- Bear Grills	drinking	piss	snake
56- Edward Scissorhads	scissoring	scissors	scissorhands
57- Egg	dancing	keg	leg
58- Bertrand Russell	binocularing	stars	binoculars
59- Justin Timberlake	folding	cemetery	grave
60- Superman	absorbing	sun	suit
61- Saruman	adivinating	Palantir	Sauron
62- Nord (Skyrim alias)	sheating	longsword	hands
63- The Hound	attacks	hound	sword
64- Darth Sidius	sparking	lightsaber	crystals
65- Jackie Chan	stunting	rags	rags
66- Spongebob Squarepants	catching	jellyfish	net
67- Angus Young	rocking	SG electric guitar	Pick of Destiny
68- Hussein	logging	logs	axe
69- Sonic	running	flag	jump
70- Orwell	turning-off	TV	controller
71- Inspector Gadget	gadgetting	gadgets	gadgets
72- George W. Bush	bombing	Bombs	bombs
73- Spiderman	webbing	Mara	webs
74- Yoda	forcing	force	Force (field)
75- Gecko	teaching	caveman	chalkboard
76- Emma Stone	lip-sticking	lips	Lip-stick
77- Gandalf the Grey	Magic (cane)	butterfly	cane
78- Gohan	genkidamaing	planet	genkidama

13

79- Qui-Gon Jinn	Mind-tricking	dealer	Mind-trick
80- Santa Claus	gifting	gifts	gifts
81- Hammer Man	bouncing	golf ball	golf club
82- Sacha Grey	seducing	cadaver	Dorian Grey
83- Hulk	smashing	house	Hulk-smash
84- Six-String Samurai	Samurai-ing	guard	samurai sword
85- Jimmi Hendrix	burning	Fender electric guitar	voodoo
86- Han Solo	blasting	storm trooper	ray-gun
87- Hermione	leviosa-ing	feather	wand
88- Hulk Hoggan	mustaches	bike	mustache
89- Hen	picking (with beak)	beggar	beak
90- Dr. No	scars	Dr. Evil	scar-tool
81- Neil Armstrong	moonwalking	spacesuit	spaceboots
92- Master Chief	pwning	Cortana	rifle
93- Nick Vujicic	flips	phone	foot
84- Napoleon Dynamite	drawing	drawing	pencil
95- Mandela	meditating	Mandela	aura
96- Obama	walking (limping)	cane	sugarcane
97- Morgan Freeman	resuscitating	zombies	white suit
98- Malcolm X	praying	Allah	Q'ran
99- Monesvol	trashing	spaghetti	Noodly appendage
00- Raptor Jesus	creates	universe	god

Other Number Systems and Encoding

1. Color System

> "Big Boys Race Our Young Girls, But Violet
> Generally Wins".

Black [1], brown [2], red [3], orange [4], yellow [5], green [6], blue [7], violet [8], grey [9], invisible (or white) [0].

The Color System can be used with a person or an object. For example, 116 can be "black [1] Arnold Shwarzenneger [16]".

2. Binary Number System

Memorizing binary numbers is the easiest of all memory feats. I encode the binary digits so they resemble decimal digits.

I use this conversion list:[4]

$$001 = 1$$
$$010 = 2$$
$$011 = 3$$
$$100 = 4$$
$$101 = 5$$
$$110 = 6$$
$$111 = 7$$

Then, I can memorize the binary digits in a PAOT System fashion. E.g., the binary number 100101110001001110001111 can be grouped as 100 101 110 001 001 110 001 111. The binary number is converted into a decimal number. In this case it converts to 45-61-16-17, which can be memorized as "Dee Dee [45] adivinating (like Saruman) [61] some weights [16]

---

[4] The correct conversion should be as follows: 001=1, 010=2 (because in binary there are only two kinds of numbers: 0 and 1. This means that the numbers from the decimal system doesn't exist), 011=3, 100=4, 101=5, and so on. For a deeper understanding I'll use here as an example a base-3 system for comparison: 001=1, 002=2, 010=3 (because in this case 3 is the base system, so there are three kinds of numbers: 0, 1 and 2), 011=4, 012=5, and so on.

with a voice-machine (that talks like Stephen Hawking) [17]".

3. Encoding Ten Decimal Numbers in a Loci

I encode ten decimal numbers or thirty binary
numbers in this fashion:

[color + P] + [A] + [O] + [color + T] = 10 decimal
digits/locus (30 binary digits/locus)

E.g.: 8794503256; 879-45-03-256; violet Qui-Gon
Jinn [8 + 79] disturbing/distracting [45] a Mario
power star [03] with a brown scissorhands [2 + 56].

Memory Palaces

1. Learning Temple

The Learning Temple[5] (LT for short) is a self-generating and artificially made memory palace in which a person can store information for classes or courses and ace exams. It resembles a roman coliseum. Every bench/step is a room filled with columns and persons. There's an allegorical statue in front of the row of benches/steps that symbolizes the name of the course. Every semester has a specific number of allegorical statues (5 courses = 5 allegorical statues). Every room has a different color and theme to make them homey and can be filled using the Roman Room Method mixed with Benavente's memory palace.

For memorizing I do the following:

1. I find my class/course.
2. In my door I place a person.
3. That person makes 20 loci.
4. I go to my nearest right-hand wall, place a person around a column and fill the loci.
5. I repeat step #4 two more times (that makes three persons around the column).

---

[5] Nicasio, Olegario (2013). *Bitácora (memory chronicle #4)*. Recovered from http://mt.artofmemory.com/blogs/olegario-nicasio/bitacora-memory-chronicle-4-4103.html

6. Then, I go to my nearest right-hand corner, place three persons around a column and fill all the loci (20 + 20 + 20 = 60 loci).

7. I go to my right-hand wall, place three persons around a column and fill the loci.

8. I go to my right-hand corner (two 'o clock), place three persons around a column and fill the loci.

9. Keep filling the columns with people until you fill all the walls and corners.

10. Place four persons around the center column and fill the loci.

11. Place four persons around the center ceiling column (same as the center column, but on the ceiling) and fill the loci.

Learning Temple

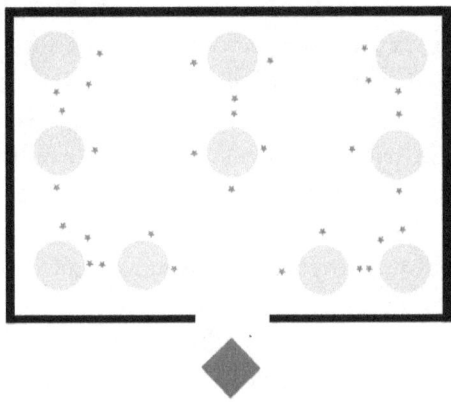

◌ = column
▢ = walls
* = person
◆ = allegorical statue

2. Artificial Memory Palace

The Artificial Memory Palace[6] (AMP for short)
is a memory palace that is made of videogame maps
instead of real world journeys. It's a good memory
palace for storing a lot of data. It gives a wide range
of possibilities, especially for people who are gamers.
I use maps from RPG games (The Elder Scrolls, The
Legend of Zelda, Megaman Battle Network). I use the
Roman Room method to generate a lot of loci. So far
I've over 13,100 loci. Just find a room (or something
similar) in your map. The door is a locus, the walls
sum up five loci, the corners sum up four loci, the center of the room is a locus and the ceiling is a locus.
That makes a total of 12 loci. If in a map you have a
100 rooms that means you have 1,200 loci. Basically
you can generate a lot of loci with few rooms!

---

[6] Nicasio, Olegario (2013). *Bitácora (memory chronicle #6)*.
Recovered from http://mt.artofmemory.com/blogs/olegario-
nicasio/bitacora-memory-chronicle-6-4173.html

References

Nicasio, Olegario (2013). *Bitácora (memory chronicle #6)*. Recovered   from http://mt.artofmemory.com/blogs/olegario-nicasio/bitacora-memory-chronicle-6-4173.html

Nicasio, Olegario (2013). *Bitácora (memory chronicle #4)*. Recovered   from http://mt.artofmemory.com/blogs/olegario-nicasio/bitacora-memory-chronicle-4-4103.html

O'Brien, Dominic (1993). *How to Develop a Perfect Memory*. Headline Book Publishing.

Octaveslur (2012). *Octaveslur's Digit System*. Recovered from http://mt.artofmemory.com/blogs/octaveslur/octaveslurs-digit-system-1889.html

Additional References

Because of the briefness of my work, I offer here additional references I have read. These references may be beneficial for both the novice and the mnemonist who wants to broaden their knowledge.

Books

Benjamin, Arthur. *Secrets of Mental Math: The Mathemagician's Guide to Lightning Calculation and Amazing Math Tricks.*

Bruno, Giordano. *Ars Reminiscendi.*

Bruno, Girdano. *Cantus Circaeus: The Incantations Of Circe Together With The Judiciary Being The Art Of Memory.*

Buzan, Tony. *The Mind Map Book: How to Use Radiant Thinking to Maximize Your Brain's Untapped Potential.*

Buzan, Tony. *Use Your Head.*

Buzan, Tony. *Use Your Memory.*

Carnegie, Dale. *The Art of Public Speaking.*

Cicero, Marcus Tullius. *On the Ideal Orator.*

Cicero, Marcus Tullius. *Rhetorica ad Herennium.*

Cooke, Ed. *Remember, Remember.*

Fauvel-Gourand, Francis. *Phreno-mnemotechny, or, The art of memory.*

Feinaigle, M. Gregor Von. *The New Art of Memory*

Foer, Joshua. *Moonwalking with Einstein: The Art and Science of Remembering Everything.*

Foster, Jonathan K. *Memory: A Very Short Introduction*

Granville, J. Mortimer. *The Secret of a Good Memory.*

Grey, R. *Dr. R. Grey's Memoria technica, or Method of artificial memory.*

Groves, W. H. *The Rational Memory.*

Konnikova, Maria. *Mastermind: How to Think Like Sherlock Holmes.*

Lavery, Michael J. *Whole Brain Power: The Fountain of Youth for the Mind and Body.*

Lorayne, Harry. *How to Develop Superpower Memory.*

Lorayne, Harry. *Page-a-Minute Memory Book.*

Luria, Alexander R. *The Mind of a Mnemonist: A Little Book about a Vast Memory.*

Miles, Pliny. *American Mnemotechny or Art of Memory.*

O'Brien, Dominic. *Quantum Memory Power: Learn to Improve Your Memory with the World Memory Champion!*

O'Brien, Dominic. *You Can Have an Amazing Memory: Lear Life-Changing Techniques and Tips from the Memory Maestro.*

Pascual, Luis Sebastián. *Lo mejor de www.mnemotecnia.es 7 años contigo.*

Pridmore, Ben. *How To Be Clever.*

Rollo, May. *The Courage to Create.*

Rowlingson, Cameron B. *Fundamentals of Memory Development.*

Shinn Boyd, Asa. *Modern Memotechny.*

Tammet, Daniel. *Born on a Blue Day: Inside the Extraordinary Mind of an Autistic Savant.*

V. Kozarenko. *GMS Manual.*

Yates, Frances A. *The Art of Memory.*

Websites

mt.artofmemory.com
www.ludism.org
www.memory-sports.com
www.world-memory-statistics.com